Work 102

非洲农场
African Farms

Gunter Pauli

[比] 冈特·鲍利　著

[哥伦] 凯瑟琳娜·巴赫　绘

姚晨辉　译

上海远东出版社

丛书编委会

主　任：田成川

副主任：何家振　闫世东　林　玉

委　员：李原原　翟致信　靳增江　史国鹏　梁雅丽

　　　　任泽林　陈　卫　薛　梅　王　岢　郑循如

　　　　彭　勇　王梦雨

特别感谢以下热心人士对童书工作的支持：

目录

Contents

非洲最大的柑橘农场过去坐落在克鲁格国家公园的外面。克鲁格国家公园是一个迷人的野生动物保护区，游客来这里可以观赏到大象、水牛、犀牛、狮子和豹子。但柑橘农场的主人意识到，自己的农场无法继续生存下去了，因为来自国外的橘子要比本地生产的便宜得多。

The largest citrus farm in Africa used to be located just outside the Kruger National Park, a wonderful game reserve where tourists come to watch elephants, buffalo, rhino, lions, and leopards. The owner of the farm realised that the farm could not survive. Oranges from overseas are much cheaper than the local ones.

非洲最大的柑橘农场

Largest citrus farm in Africa

我们不得不关闭农场

We have to close down the farm

农场主基斯说："在世界各地，柑橘类水果的价格都跌得这么厉害，我们一点钱都赚不到了，这样下去我们不得不关闭农场。"

罗伊索——他的朋友，也是邻居，劝说道："专家们宣称，在这个经济全球化的时代，你只有降低成本，裁减员工，更好地利用灌溉水，使用化学品，才能生存下去。"

"The price of citrus fruit has dropped so much worldwide, that we cannot make any money anymore.
So, we have to close down the farm,"
says Khethi, the farm owner.

"Experts claim that you have to cut costs, lay off staff, use more irrigation water, and use chemicals to make ends meet in this globalized economy," argues Loyiso, his friend and neighbour.

"我们的橘子不可能像世界上其他地方的那样便宜。我们是一个合作社，所有的成员都是业主，不可能将任何人解雇。"

"你不能老想着种出更便宜的橘子，你应该试试利用现有的橘子赚更多的钱。" 罗伊索建议道。

"怎么改变现在的柑橘经营模式呢？没有人知道，更不要说了解。每个人都告诉我，我们能做的就是用更便宜的价格卖掉它们。"

"We cannot be as cheap as the others around the world. As we are a cooperative, and all members are owners, it's impossible to kick out anyone."

"You should not try to grow cheaper oranges. You should try to earn more with the ones you already have," suggests Loyiso.

"No one knows or even understands how you can do more with an orange than we are doing now. And everyone tells me that we have to sell them at a cheaper price."

利用现有的橘子赚更多的钱

Earn more with the ones you already have

你一年四季都可以供应橘子吗？

Do you have oranges all year around?

"降低成本，这个主意不错，可是现在已经有这么多的人失业，你不能简单地把工人辞退了事，你必须对人民和社会承担一份责任。"

"要这样我们就没有成功的希望了！" 基斯说。

"别放弃呀，我们一起想想可行的方法。你一年四季都可以供应橘子吗？"

"It's not a bad idea to reduce the costs, but when there are already so many people without jobs, you cannot simply kick your workers out. You have a responsibility towards people and the community."

"Then there's no way we could be successful!" says Khethi.

"Don't give up. Let's find a way to make it work. Do you have oranges all year around?"

"差不多吧，我们林波波省的气候好得很，这里可是阳光明媚的南非啊！"

"就是嘛，既然你有那么多橘子，你可以榨果汁啊。"

"这简单！而且我们可以向克鲁格国家公园甚至周围所有的旅馆提供果汁。"

"Nearly. We have a great climate here in Limpopo. This is sunny South Africa!"

"So, if you have oranges, you can make juice."

"That's easy! And we can deliver it to all the lodges in and around the Kruger Park."

你可以榨果汁啊

you can make juice

你可以生产可生物降解的肥皂

you can make biodegradable soaps

"榨果汁，就会有残留的
果皮。对果皮进行蒸煮，你就可以生产
可生物降解的肥皂。"
"你要我们这些种柑橘的农民去生产和销售肥皂？"
基斯问道。
"不，我认为你应该为那些旅馆提供洗衣服务。因为
你的农场制作的化学品是天然的，它们能够
将废水转化为灌溉用水。"

"And when you make juice, you'll have
leftover peels. You can steam the peels to
make biodegradable soaps."

"Do you want us, the orange farmers, to produce
and sell soap?" asks Khethi.

"No, I think you should offer laundry services to
the lodges. As the chemicals coming from your
farm are natural, they will be able to turn their
waste water into irrigation water."

"嗯，果树还需要修剪，我们可以用剪下来的树枝种植蘑菇。"

"这个主意真棒！"罗伊索说，"蘑菇菌糠也是很好的动物饲料。我还在想，旅馆自助餐厅里有很多食物都是做点缀用的，你可以将这些剩菜要来作为猪饲料。"

"And as fruit trees need pruning, we could use the prunings to grow mushrooms on."

"What a good idea!" says Loyiso. "Mushroom substrate is also good feed for animals. I was just thinking, as so much of what you see at the buffet at these lodges is only for decoration, you could ask them if you could use those leftovers as pig feed."

种植蘑菇

Grow mushrooms

用废弃物生产沼气

Waste to produce biogas

"我觉得旅馆老板会很乐意把他们的废弃物给我们去生产沼气。"

"这就意味着，多亏了这些橘子，你可以比仅仅卖水果赚更多的钱。"

"I think the lodge owners would happily supply us with their waste to produce biogas."

"That means that thanks to the oranges, you will be able to make more money than when you only export the fruit."

"现在我们不用再担心国际市场上橘子的价格啦，我们可以创造更多以前没有想到的就业机会。这才是我喜欢的那种经济模式！"

……这仅仅是开始！……

"Now we don't have to worry about the price of oranges on the world market and we can create more jobs than we ever imagined. That is the kind of economy I like!"

... AND IT HAS ONLY JUST BEGUN!...

……这仅仅是开始！……

… AND IT HAS ONLY JUST BEGUN! …

Did You Know?

你知道吗？

The first orange farms were started in 2500 BC in China. The sweet oranges originated from India and were introduced to the world by the Portuguese. The bitter orange originated from Persia, and was first introduced to Italy.

公元前2500年，世界上第一个柑橘农场出现在中国。甜橘起源于印度，被葡萄牙人传播到世界各地。苦橘原产波斯，最先被出口到意大利。

Arab, Portuguese, Spanish, and Dutch sailors planted orange trees along trade routes to ensure a supply of fruit to prevent scurvy.

阿拉伯、葡萄牙、西班牙和荷兰的水手沿着他们的贸易路线种植橘子树，以确保水果供应，预防坏血病。

Orange trees are the most cultivated trees in the world and are widely grown in tropical and subtropical climates. Brazil is the largest producer in the world.

橘子树是世界上最常见的树木，广泛种植于热带和亚热带地区。巴西是世界上最大的柑橘生产国。

The word "orange" originates from the Arabic nãranj, and the "n" was dropped in many languages as the article "an" already contains an "n". In French, for example, saying une norange sounded like there was one "n" too many.

单词"橘子"（orange）来源于阿拉伯语nãranj，"n"这个字母后来在许多语言里都被省略了，因为冠词"an"中已经包含了一个"n"。比如在法语中，如果把"一只橘子"说成"une norange"，听上去就显得其中的"n"太多了。

The pH of an orange can be as low as 2.9 and as high as 4.0. This is as acidic as a cup of coffee or a cola drink.

橘子的pH值在2.9和4.0之间，相当于一杯咖啡或者可乐的酸性。

In Florida (USA) and Brazil, oranges are harvested with a canopy-shaking machine. These two places are the largest producers of frozen concentrated orange juice.

在美国的佛罗里达州和巴西，人们会用一种树冠振动机来采摘橘子。这两个地方也是冷冻浓缩橘汁的最大生产地。

果汁通过货船运往世界各地，这种船被称为果汁油轮，每艘轮船可以运送3.5万吨果汁。

Juice is shipped around the globe by cargo boats, called fruit juice tankers, that can transport 35,000 tonnes of juice in one trip.

橘子皮中含有丰富的油脂，可以用来生产油漆和清洁产品，其价格随着橘子的季节价格而波动。

Orange peel is rich in oil that can be used in paints and cleaning products. The price fluctuates with the seasonal price of oranges.

Would you be able to tell friends who have worked with you for years and co-owns your company that they have to lose their jobs?

你做得到吗，告诉与你共事多年的朋友和公司的合伙人，他们将要被解雇？

如果每升橘皮油比同样数量的橙汁可以多赚10倍的钱，你会只生产橘皮油呢，还是会两者同时进行？

If you can earn 10 times more money per litre of citrus peel oil than with the same amount of orange juice, would you rather work only with peels, or with both?

If you had fruit trees, would you only want to do business when the fruit are in season, or would you like to find a way to do business all year around?

如果你有果树，你是只想做当季水果的生意呢，还是说你会设法找到一种办法一年四季都可以做生意？

将水果和食品作为一种美丽的装饰，可以吸引人们多吃一点吗？或者，人们宁愿把被丢弃的水果用作猪饲料？

Are fruit and food a fine form of decoration that will entice people to eat more, or would people rather be inspired by the fact that the waste from fruit serve as feed for pigs?

Squeeze the juice from some oranges. Now squeeze the peels. How much oil can you get from the peels? It may be easier to eat just eat the peels. Do make sure the oranges were organically grown before you do this. You may not want to eat the peel by itself, so the question is how you can use the peel in your food so that you will be able to get health benefits from it. Grate it onto your salad or desert, add it to you mom's coffee, or cover it in dark chocolate. Who likes the taste? Is anyone willing to eat only the peel?

　　从几个橘子中挤一些橘子汁。然后再挤一挤橘子皮,你从橘子皮中可以挤出多少橘皮油?更容易的吃法是单独吃皮,但你这样做之前一定要确保橘子是有机种植的。你可能不想单独吃橘子皮,那问题来了,如何将橘子皮加进食物中,以便你能够获得橘子皮中对健康有益的成分?可以将橘皮碾碎放到你的沙拉中,添加到妈妈的咖啡里,或撒在黑巧克力上面。有人喜欢这种味道吗?有人愿意单独吃橘皮吗?

学科知识
Academic Knowledge

生物学	坏血病是一种因缺乏维生素C造成的疾病，症状包括疲劳、海绵状牙龈出血等；橘子通过授粉或单性结实进行繁殖；橘子是柚子和柑的杂交品种；橘子可以混种繁殖，即很容易产生杂交品种；橘子树可以嫁接，即将一种植物的组织插入另一种植物中（接合），以获得新的树木；橘子树是硬木树，因此其所有的木材废料都适合蘑菇种植；橘子皮包含的纤维比果肉多4倍；橘子皮30%的成分是果胶，果胶是一种天然凝胶。
化 学	酸度低的橘子容易腐烂；橙子含有类胡萝卜素和类黄酮；橘子即使在采摘之后还能呼吸，放出氧、二氧化碳和乙烯。
物 理	水溶性和水稀释性的区别；冰凉的水果没有温热的水果榨出的果汁多；中等大小的橘子含有更多的果汁；湿度过大会加速橘子腐烂。
工程学	通过蒸馏将油从果皮中分离出来可以生产出食品级的油；巴氏杀菌法可以确保果汁全年的供应，使生产与销售不再有季节限制；浓缩果汁意味着将果汁中的水蒸发掉，需要加水复原。
经济学	企业的法律架构：有限责任、股份以及合伙；橘子和橘汁的价格由商品市场所决定，有目前的和预期的价格；为什么市场上新鲜水果的价格比果汁贵得多；橘子（像大多数水果和蔬菜一样）有季节性，因此价格每年波动。
伦理学	为什么我们会花更多的钱去买外表好看的水果，而不是买具有相同品质和口味却不好看的水果；超市的政策是不卖外表不"完美"的水果和蔬菜；参观野生动物保护区的人往往生活奢华，享受进口食品，但大部分费用都付给了经营者，而不是当地居民。
历 史	公元前314年的中国文献里已经提到了甜橘；哥伦布将橘子的种子带到了加勒比海地区。
地 理	林波波省是南非的一个省；南非的克鲁格国家公园成立于1926年，其园区的一部分位于姆普马兰加省，一部分位于林波波省。
数 学	虽然柠檬中d-柠檬烯的含量是橘子的约两倍，橘子的体积却大得多。
生活方式	新鲜的橘汁已成为很多人每日早餐的一部分。一杯橘汁足以提供每日所需的维生素C。
社会学	为子孙后代考虑，我们必须具有保护野生动物园区的意识。
心理学	需要有一个专门的人员负责解雇员工，尤其当他们是你多年的朋友和同事时。
系统论	果园经受的价格波动超出了果园主人的控制，为了实现可持续发展，必须创造其他的收入来源，额外的收入要从当地经济中产生，以保持稳定性和增强适应力。

情感智慧
Emotional Intelligence

主　人

果园的主人处于绝望之中，感到别无选择，只有关闭果园。他意识到了合作社的局限性，却坚持不能将朋友解雇。他很失望，没有专家可以为他提供解决方案。水果的价格波动让他不知所措，他明白自己的成本太高，却找不到降低成本应对国际市场竞争的办法。他认为自己走投无路了。不过，他很喜欢家乡的气候，并为自己的国家感到骄傲。在一次和邻居的开放性的对话中，他慢慢开始了解到一些可能的解决方案，并高兴地从中受到了启发。他渴望了解更多信息，将他所拥有的知识和似乎毫无关联的知识结合起来，明确如何才能自强自立，且不必解雇任何员工。

邻　居

邻居很现实，尽管他知道专家的建议，却没有就这个话题展开辩论，而是试图找到方法摆脱看似走投无路的情况。他指出了果园主人的社会责任，给果园主人打气，认为他不应该放弃，并通过与他交谈，寻求解决方案。他提供了一条捷径：制作橘汁。果园主人接受了这个想法。他得到鼓励，又提出了第二个不太为人所知的商业点子：将果皮变成肥皂。邻居继续寻找更高的价值实现途径，他建议不要卖肥皂，而是提供洗衣服务，并因此获得额外的灌溉用水利益。当邻居发现果园主人的心态转变之后，他提出了更多的想法，可以让农场赚更多的钱，让大家对未来充满希望。

艺术
The Arts

让我们利用橘子来进行艺术创作。柑橘是一种果皮较厚的水果，其果皮可以被切割和扭曲成复杂的形状。看看你能不能用橘子皮拼成大象和蜗牛的形状。这比仅仅吃掉水果，然后扔掉果皮有趣多了！你会发现艺术太有趣了，除了享受水果的美味之外，你也可以用橘子创造美丽的艺术品。

思维拓展
Systems: Making the Connections

从事水果商品化生产的公司很难在全球市场上进行竞争，除非它们实现了非常高的规模经济。橘子也不例外。最初热带和亚热带地区贸易航道沿线的所有国家都引进了橘子，随着时间的流逝，橘子的生产集中到了低成本的地区（如巴西），或具有高水平的自动化和科学化种植管理的地区。而那些没有低成本优势或科学化生产企业的国家应该怎么做呢？商业顾问的传统建议总是始终不变的：增加经济规模，提高生产效率，减少员工人数，充分利用灌溉和遗传学。另一种选择是倒闭，寻找替代产业。上述选择虽然有一定的逻辑和道理，但现实的困难在于解雇员工和关闭现有的企业并不容易，这将降低地区经济收入，对整个社区，从本地面包店到城镇的税收收入等都产生影响。如果组织的结构形式是合作社性质的，所有的员工同时也是股东，那么在不造成紧张和压力以及破坏社会关系的情况下，裁员几乎是不可能的。最好的方法是评估该地区的经济活动，看看如何将果园的生产更好地融入该地区的现有商业。另一方面，一旦地方经济的定位明确之后，可以很快找到促进本地消费的方法。接下来的产品和服务都将和橘子及其加工过程相关联，还包括客户网络的出现、建立社区等。未来将发生翻天覆地的变化，将有更多的钱在当地经济中流通，提供更多的购买力，增强就业市场。最重要的是，使柑橘种植业不再受世界市场上价格波动的影响。

动手能力
Capacity to Implement

如果由你负责一家销售橘子汁的商店，你打算怎样常年为客户服务？如果你只卖橘子，那么你将不得不从海外进口一些，这就需要存储空间，因为橘子在采摘后只能保存一段时间。查看你需要从什么地方进口橘子，以保证一年12个月都有备货。另一种选择是只在当季销售本地的橘子，在橘子过季之后寻找其他种类的水果制作果汁。你将会使用哪些水果来保证你全年的业务运营？你打算如何说服客户，他们仍然会得到日常所需的新鲜维生素C？研究你所有可能的选项，制定一个计划，并向你的朋友和你的父母进行介绍。

故事灵感来自
This Fable Is Inspired by

路易吉·比斯塔吉诺
Luigi Bistagnino

　　路易吉·比斯塔吉诺读书时的专业是建筑学，目前他是都灵理工大学建筑和设计学院的教授。他在将学术课程（以前仅限于产品和工艺设计）转变为系统设计方面进行了探索。他与"慢食运动组织"密切合作，致力于将系统设计应用于所有的经济活动中。他特别擅长将这些新概念应用到农业和食品加工行业中，并且积累了丰富的研究案例。比斯塔吉诺教授论证了如何将种植与水果和蔬菜的加工相结合，以促进地方经济发展和创业。创造的价值将转化为更多的财富、工作岗位、适应性、生物多样性，提高文化认同感。如果尝试将当地的农业和加工业融入全球经济之中，将获得比以往任何时候都更好的生活质量。

图书在版编目(CIP)数据

冈特生态童书.第三辑修订版:全36册:汉英对照 /
(比)冈特·鲍利著;(哥伦)凯瑟琳娜·巴赫绘;
何家振等译.—上海:上海远东出版社,2022
书名原文:Gunter's Fables
ISBN 978-7-5476-1850-9

Ⅰ.①冈… Ⅱ.①冈… ②凯… ③何… Ⅲ.①生态环
境–环境保护–儿童读物—汉、英 Ⅳ.①X171.1-49

中国版本图书馆CIP数据核字(2022)第163904号
著作权合同登记号图字09-2022-0637号

策　　划　张　蓉
责任编辑　程云琦
封面设计　魏　来李　廉

冈特生态童书
非洲农场

[比]冈特·鲍利　著
[哥伦]凯瑟琳娜·巴赫　绘

姚晨辉　译

记得要和身边的小朋友分享环保知识哦!
八喜冰淇淋祝你成为环保小使者!